Measuring Performance and Benchmarking Project Management at the Department of Energy

Committee for Oversight and Assessment of
U.S. Department of Energy Project Management

Board on Infrastructure and the Constructed Environment

Division on Engineering and Physical Sciences

NATIONAL RESEARCH COUNCIL
OF THE NATIONAL ACADEMIES

THE NATIONAL ACADEMIES PRESS
WASHINGTON, D.C.
www.nap.edu

THE NATIONAL ACADEMIES PRESS 500 Fifth Street, N.W. Washington, DC 20001

NOTICE: The project that is the subject of this report was approved by the Governing Board of the National Research Council, whose members are drawn from the councils of the National Academy of Sciences, the National Academy of Engineering, and the Institute of Medicine. The members of the committee responsible for the report were chosen for their special competences and with regard for appropriate balance.

This study was supported by Contract Number DEAM01-99PO8006 between the U.S. Department of Energy and the National Academy of Sciences. Any opinions, findings, conclusions, or recommendations expressed in this publication are those of the author(s) and do not necessarily reflect the views of the organizations or agencies that provided support for the project.

International Standard Book Number 0-309-09708-8

Additional copies of this report are available from the National Academies Press, 500 Fifth Street, N.W., Lockbox 285, Washington, DC 20055; (800) 624-6242 or (202) 334-3313 (in the Washington metropolitan area); Internet, http://www.nap.edu

Printed in the United States of America

THE NATIONAL ACADEMIES

Advisers to the Nation on Science, Engineering, and Medicine

The **National Academy of Sciences** is a private, nonprofit, self-perpetuating society of distinguished scholars engaged in scientific and engineering research, dedicated to the furtherance of science and technology and to their use for the general welfare. Upon the authority of the charter granted to it by the Congress in 1863, the Academy has a mandate that requires it to advise the federal government on scientific and technical matters. Dr. Bruce M. Alberts is president of the National Academy of Sciences.

The **National Academy of Engineering** was established in 1964, under the charter of the National Academy of Sciences, as a parallel organization of outstanding engineers. It is autonomous in its administration and in the selection of its members, sharing with the National Academy of Sciences the responsibility for advising the federal government. The National Academy of Engineering also sponsors engineering programs aimed at meeting national needs, encourages education and research, and recognizes the superior achievements of engineers. Dr. Wm. A. Wulf is president of the National Academy of Engineering.

The **Institute of Medicine** was established in 1970 by the National Academy of Sciences to secure the services of eminent members of appropriate professions in the examination of policy matters pertaining to the health of the public. The Institute acts under the responsibility given to the National Academy of Sciences by its congressional charter to be an adviser to the federal government and, upon its own initiative, to identify issues of medical care, research, and education. Dr. Harvey V. Fineberg is president of the Institute of Medicine.

The **National Research Council** was organized by the National Academy of Sciences in 1916 to associate the broad community of science and technology with the Academy's purposes of furthering knowledge and advising the federal government. Functioning in accordance with general policies determined by the Academy, the Council has become the principal operating agency of both the National Academy of Sciences and the National Academy of Engineering in providing services to the government, the public, and the scientific and engineering communities. The Council is administered jointly by both Academies and the Institute of Medicine. Dr. Bruce M. Alberts and Dr. Wm. A. Wulf are chair and vice chair, respectively, of the National Research Council.

www.national-academies.org

COMMITTEE FOR OVERSIGHT AND ASSESSMENT OF U.S. DEPARTMENT OF ENERGY PROJECT MANAGEMENT

LLOYD A. DUSCHA, *Chair*, U.S. Army Corps of Engineers (retired), Reston, Virginia
DON JEFFREY BOSTOCK, Lockheed Martin Energy Systems (retired), Oak Ridge, Tennessee
ALLAN V. BURMAN, Jefferson Solutions, Washington, D.C.
G. BRIAN ESTES, Consulting Engineer, Williamsburg, Virginia
STEVEN L. FAHRENKROG, Project Management Institute, Newton Square, Pennsylvania (resigned January 2005)
DAVID N. FORD, Texas A&M University, College Station
THEODORE C. KENNEDY, BE&K, Inc., Birmingham, Alabama
STEPHEN R. THOMAS, Construction Industry Institute, Austin, Texas

Staff

LYNDA L. STANLEY, Director, Board on Infrastructure and the Constructed Environment
RICHARD G. LITTLE, Director, Board on Infrastructure and the Constructed Environment (through December 1, 2004)
MICHAEL D. COHN, Program Officer
DANA CAINES, Financial Associate
PAT WILLIAMS, Senior Project Assistant

BOARD ON INFRASTRUCTURE AND THE CONSTRUCTED ENVIRONMENT

PAUL GILBERT, *Chair*, Parsons Brinckerhoff Quade & Douglas, Seattle, Washington
MASSOUD AMIN, University of Minnesota, Minneapolis
RACHEL DAVIDSON, Cornell University, Ithaca, New York
REGINALD DesROCHES, Georgia Institute of Technology, Atlanta
DENNIS DUNNE, Consultant, Scottsdale, Arizona
PAUL FISETTE, University of Massachusetts, Amherst
LUCIA GARSYS, Hillsborough County, Florida
WILLIAM HANSMIRE, Parsons Brinckerhoff Quade & Douglas, San Francisco, California
HENRY HATCH, U.S. Army Corps of Engineers (retired), Oakton, Virginia
AMY HELLING, Georgia State University, Atlanta
THEODORE C. KENNEDY, BE&K, Inc.
SUE McNEIL, University of Chicago, Illinois
DEREK PARKER, Anshen+Allen, San Francisco, California
HENRY SCHWARTZ, JR., Washington University, St. Louis, Missouri
DAVID SKIVEN, General Motors Corporation, Detroit, Michigan
MICHAEL STEGMAN, University of North Carolina, Chapel Hill
WILLIAM WALLACE, Rensselaer Polytechnic Institute, Troy, New York
CRAIG ZIMRING, Georgia Institute of Technology, Atlanta

Staff

LYNDA STANLEY, Director, Board on Infrastructure and the Constructed Environment
RICHARD G. LITTLE, Director, Board on Infrastructure and the Constructed Environment (through December 1, 2004)
MICHAEL D. COHN, Program Officer
DANA CAINES, Financial Associate
PAT WILLIAMS, Senior Project Assistant

Acknowledgment of Reviewers

This report has been reviewed in draft form by individuals chosen for their diverse perspectives and technical expertise, in accordance with procedures approved by the NRC's Report Review Committee. The purpose of this independent review is to provide candid and critical comments that will assist the institution in making its published report as sound as possible and to ensure that the report meets institutional standards for objectivity, evidence, and responsiveness to the study charge. The review comments and draft manuscript remain confidential to protect the integrity of the deliberative process. We wish to thank the following individuals for their review of this report:

Philip R. Clark, Nuclear Corporation (retired),
Henry J. Hatch, U.S. Army Corps of Engineers (retired),
Peter Marshall, Burns & Roe Services,
Julia Melkers, University of Illinois, and
James M. Tien, Rensselaer Polytechnic Institute.

Although the reviewers listed above have provided many constructive comments and suggestions, they were not asked to endorse the conclusions or recommendations, nor did they see the final draft of the report before its release. The review of this report was overseen by Gerald E. Galloway, University of Maryland. Appointed by the National Research Council, he was responsible for making certain that an independent examination of this report was carried out in accordance with institutional procedures and that all review comments were carefully considered. Responsibility for the final content of this report rests entirely with the authoring committee and the institution.

Contents

EXECUTIVE SUMMARY 1

1 INTRODUCTION 3
Background, 3
Organization of This Report, 4
References, 5

2 PROJECT MANAGEMENT PERFORMANCE MEASURES 7
Control System for Process Improvement, 7
Input, Process, Output, and Outcome Measures, 8
Selecting Effective Performance Measures, 10
References, 11

3 THE BENCHMARKING PROCESS 21
Introduction, 21
Benchmarking Roadmap, 22
External Versus Internal Benchmarking, 22
Implementation of Benchmarking, 24
References, 25

4 IMPLEMENTATION 27
Introduction, 27
Development of a Performance Measurement and Benchmarking System, 27
Data Collection and Validation, 28
Benchmarking Readiness, 29
References, 29

APPENDIXES

A NRC's DOE Project Management Assessment Reports 33
B Statement of Task 35
C Biographies of Committee Members 37
D Discussions with DOE Personnel 41

Acronyms and Abbreviations

CD-0	critical decision 0, approval of mission need
CD-1	critical decision 1, approval of system requirements and alternatives
CD-2	critical decision 2, approval of project baseline
CD-3	critical decision 3, authorization to complete implementation
CD-4	critical decision 4, approval of project completion and transition to operations
CII	Construction Industry Institute
DMAIC	define, measure, analyze, improve, and control
DOE	U.S. Department of Energy
EIR	external independent review
EM-PDRI	Office of Environmental Management Project Definition Rating Index
ESAAB	Energy System Acquisition Advisory Board
EVMS	Earned Value Management System
GAO	Government Accountability Office
GPRA	Government Performance and Results Act of 1993
IPR	independent project review
IPT	integrated project team
OECM	Office of Engineering and Construction Management
OMB	Office of Management and Budget
PARS	Project Analysis and Reporting System
PSO	program secretarial office

Executive Summary

The National Research Council (NRC) completed a planned 3-year review and assessment of the U.S. Department of Energy's project management with publication of *Progress in Improving Project Management at the Department of Energy: 2003 Assessment* (NRC, 2004). In that final assessment report, the Committee for Oversight and Assessment of U.S. Department of Energy Project Management made the following observation:

> DOE does not have a uniform set of objective measures for assessing the quality of project management. The lack of objective measures or even reliable historic project data makes it difficult to assess progress in improving project management. It also makes it difficult to build confidence within GAO, Congress, OMB, and the public in the department's ability to manage the money it spends on its projects. Evidence continues to be anecdotal rather than objective, quantitative, and verifiable. The absence of objective performance measures prevents the identification of best practices and impedes widespread improvement in project management throughout the agency. (NRC, 2004, pp. 31-32)

The Department of Energy responded to the NRC report by forming an internal task group led by the Office of Engineering and Construction Management (OECM) to develop performance measures and benchmarking procedures and asked the NRC to provide additional assistance to guide this effort. This report, therefore, does not contain new findings or recommendations. The advice and guidance presented are intended to help DOE develop and implement effective performance measures and an effective benchmarking program for project management. The information and guidance should be viewed not as the final work but rather as a first step toward development of a viable methodology to suit the needs and goals of DOE.

For a performance measurement and benchmarking system to be effective, it should be crafted to fill multiple organizational needs, carry the imprimatur of the users, and be accepted at all levels of the organization.

The committee suggests 30 possible performance measures in four sets:

- *Project-level input/process measures.* Assess the resources provided to deliver an individual project and the management of the project against standard procedures.
- *Project-level output/outcome measures.* Assess the cost and schedule variables of an individual project and the degree to which the project achieves the stated objectives.
- *Program- and department-level input/process measures.* Assess the total resources provided for all projects within a program or department and the degree to which program- and department-wide goals for projects and their management are met.
- *Program- and department-level output/outcome measures.* Assess overall project performance and the effectiveness of completed projects in supporting program and department missions.

The value of an individual performance measure is limited, but, combined, the measures provide a robust assessment of the quality of project management for individual projects and programs. If applied consistently over time and used for internal and external benchmarking, the measures will provide the information needed for day-to-day management and long-term process improvement. It remains the responsibility of DOE to select and use the measures that work best for it and its program offices.

For performance measures to have meaning and provide useful information, it is necessary to establish comparisons. The comparisons may evaluate progress in achieving given goals or targets, assess trends in performance over time, or weigh the performance of one organization against another.

Benchmarking is an integral part of process improvement that provides a mechanism for making comparisons of project and program performance internally and externally. To be successful, benchmarking should be implemented as a structured, systematic process based on an understanding of critical success factors. Benchmarking can be applied during various phases of a project for different purposes. When applied early on, such as at project authorization, it can be used to identify characteristics that may be associated with potential future problems and to identify aspects of project management (e.g., risk management) that need special attention to ensure project success. When applied during project execution, it can serve as a management tool to guide project decisions. Postproject benchmarking is usually used to assess performance of a project delivery system and to establish benchmarks for future comparisons.

The benchmarking process described in this report involves nine activities:

- Determine what to benchmark,
- Define the measures,
- Develop data collection methodology,
- Collect data,
- Identify deficiencies in the use of best practices and project management performance,
- Identify reasons for deficiencies (root causes),
- Develop an action plan (select best practices to reduce deficiencies),
- Integrate best practices into the project delivery process, and
- Institutionalize benchmarking as part of a continuous improvement program.

This report primarily addresses the first four of the nine steps. The remaining five steps define the essence and purpose of benchmarking, which is to continuously improve project management. For the effort to yield any benefits, it is essential that DOE establish a means for implementing the five steps after the collection of data and a system for continuous feedback and evaluation.

Measuring performance and benchmarking should be viewed as a routine, integral part of project management processes rather than a separate function. This requires that advocacy be built into the system. The most difficult step is establishing an organizational culture that is ready to assess, compare, and analyze performance and to adopt best practices used by others when they are identified. This requires an organizational commitment to continuous improvement, acceptance of new ideas, and open communication and cooperation at all levels of the organization.

Development of the necessary organizational readiness for benchmarking can be facilitated by taking incremental steps, starting with a limited number of measures and internal benchmarking within a program, then expanding the effort to include more diverse measures and comparisons with other programs. The objective over time should be to develop a full set of measures and to benchmark both internally and externally with organizations in other government agencies and private industry.

1
Introduction

BACKGROUND

In response to a directive from the Committee of the Conference on Energy and Water Development of the 105th Congress (U.S. Congress, 1999), DOE requested that the National Research Council (NRC) appoint a committee to review and assess the progress made by the department in improving its project management practices. The NRC appointed the Committee for Oversight and Assessment of U.S. Department of Energy Project Management, under the auspices of the Board on Infrastructure and the Constructed Environment (BICE), to undertake the oversight and assessment of DOE project management. The committee completed its planned 3-year review with publication of *Progress in Improving Project Management at the Department of Energy: 2003 Assessment* (NRC, 2004a). (See Appendix A for a complete list of NRC reports on DOE project management.) In its final assessment the committee made the following observations:

> DOE does not have a uniform set of objective measures for assessing the quality of project management. The lack of objective measures or even reliable historic project data makes it difficult to assess progress in improving project management. It also makes it difficult to build confidence within GAO, Congress, OMB, and the public in the department's ability to manage the money it spends on its projects. Evidence continues to be anecdotal rather than objective, quantitative, and verifiable. The absence of objective performance measures prevents the identification of best practices and impedes widespread improvement in project management throughout the agency. (NRC, 2004, pp. 31-32)

In addition, it observed that

> benchmarking performance and management processes throughout a project's life cycle and from different perspectives can provide a basis for a measure of improvement of project management procedures. Both internal and external benchmarking perspectives are useful and should be a regular part of DOE benchmarking procedures. (NRC, 2004, p. 33)

The Department of Energy responded to the NRC report by forming an internal task group, led by the Office of Engineering and Construction Management (OECM), to develop performance measures and benchmarking procedures and asked the NRC to provide additional assistance to guide this effort. For this purpose, the NRC's Board on Infrastructure and the Constructed Environment reconvened the Committee for Oversight and Assessment of U.S. Department of Energy Project Management. Six members of the earlier committee were involved in this study, together with two new members. The committee has experience in academic, government, and industrial settings and extensive knowledge of project management, process improvement, performance measurement, and benchmarking. (See Appendix B for the committee's statement of task and Appendix C for biographies of the committee members.)

The committee reviewed ongoing DOE benchmarking efforts, conducted brainstorming sessions, and convened discussions with OECM and DOE program office representatives. (See Appendix D for a list of participants in the committee briefings and discussions.) The committee appreciates the valuable contributions of the DOE personnel who participated in the committee sessions. The committee urges all program offices and project directors to become actively engaged in DOE's efforts to develop and implement effective performance measures and benchmarking processes.

In past reports, the committee emphasized the need for collecting data to both assess project performance and support project management decisions (NRC, 2001, 2003, 2004a). DOE has implemented the Project Analysis and Reporting System (PARS). However, as noted in the 2003 assessment, current project performance data available in PARS are not sufficient for assessing DOE project management because of problems with inconsistent data and the lack of historical trend data. PARS was created to collect high-level data for senior management review. The committee has found that such systems are more likely to function well if the collected data are also used at the project level. OECM recognizes these problems and opportunities and is currently working to improve the reliability and applicability of PARS; therefore, these issues are not addressed further in this report. When PARS is able to provide accurate and timely data, and when DOE can accurately track these data over time, it will be an important part of performance assessment and, if acted on, will contribute to process improvement.

In 2001, the committee conducted a government/industry forum on the owner's role in project management (NRC, 2002). The case studies presented at the forum told of companies experiencing a marked improvement in project success. Their success was attributed at least in part to a commitment to process improvement through consistent use of company-wide performance measures. They achieved success by focusing on the project management process, making performance measures part of the core values of the company and assuring that the measures were consistent throughout the organization. The companies developed a corporate language for defining and controlling the project management process and developed performance measures to guide process improvement. The performance measures and benchmarking procedures discussed in this report are tools that can help DOE make performance measurement part of its core values and processes.

The committee notes that process improvement can be assessed by analysis of performance trends for projects initiated over a period of time (NRC, 2004a). Although it may be years before benchmarking produces observable significant results throughout the DOE, it is essential to long-term performance improvement and should start now. In addition to long-term benefits, an investment in a benchmarking process provides immediate feedback to project directors as they assess their management procedures and the information they use to make decisions.

The NRC recently published the report *Intelligent Sustainment and Renewal of Department of Energy Facilities and Infrastructure* (NRC, 2004b), which discusses DOE's need for effective performance measures and benchmarking for managing facilities after projects are completed. The report suggests an asset life-cycle approach that addresses the integration of management decisions from project inception through facility disposal. To achieve this integration, metrics and benchmarking for all phases need to be integrated. The facilities and infrastructure report included a detailed set of performance measures and approaches to benchmarking that should be considered when developing a system for project management.

The information and advice presented in this report are intended to help DOE to develop and implement effective performance measures and an effective benchmarking program for project management. However, this is only a beginning, and an ongoing and consistent commitment to continuously refine and implement the process will be needed. Such efforts have worked in the private sector and at other federal agencies (NPR, 1997), and they can be made to work for DOE.

ORGANIZATION OF THIS REPORT

Following the background information on recent NRC oversight and assessment of DOE project management given in Chapter 1, Chapter 2, "Project Management Performance Measures," describes performance measures as part of an ongoing system for controlling process improvement and in terms of a paradigm that includes input, process, outcome, and output measures. Chapter 2 also describes the characteristics of effective measures and suggests measures for use at the project, program, and departmental levels.

Chapter 3, "The Benchmarking Process," describes the benchmarking process developed by the Construction Industry Institute to assess and improve construction project management. The chapter includes a roadmap for implementing a benchmarking system and the critical factors affecting the performance of these systems. The characteristics of internal versus external benchmarking and approaches to quality control are also discussed. Chapter 4, "Implementation," provides a perspective on how the information in this report can be used by DOE to implement an effective performance measurement and benchmarking system.

As mentioned above, this report includes four appendixes: a list of NRC reports on DOE project management, the statement of task, biographies of committee members, and a list of DOE personnel who participated in discussions with the committee.

REFERENCES

NPR (National Partnership for Reinventing Government Reports, formerly the National Performance Review). 1997. *Serving the American Public: Best Practices in Performance Measurement.* Available online at http://govinfo.library.unt.edu/npr/library/papers/benchmrk/nprbook.html. Accessed March 14, 2005.

NRC (National Research Council). 2001. *Progress in Improving Project Management at the Department of Energy, 2001 Assessment*. Washington, D.C.: National Academy Press.

NRC. 2002. *Proceedings of Government/Industry Forum: The Owner's Role in Project Management and Preproject Planning*. Washington, D.C.: National Academy Press.

NRC. 2003. *Progress in Improving Project Management at the Department of Energy, 2002 Assessment*. Washington, D.C.: The National Academies Press.

NRC. 2004a. *Progress in Improving Project Management at the Department of Energy, 2003 Assessment*. Washington, D.C.: The National Academies Press.

NRC. 2004b. *Intelligent Sustainment and Renewal of Department of Energy Facilities and Infrastructure.* Washington, D.C.: The National Academies Press.

U.S. Congress. 1999. House of Representatives, Energy and Water Appropriations Bill, 2000. HR 106-253. Washington, D.C.: Government Printing Office.

2
Project Management Performance Measures

CONTROL SYSTEM FOR PROCESS IMPROVEMENT

The purpose of performance measurement is to help organizations understand how decision-making processes or practices led to success or failure in the past and how that understanding can lead to future improvements. Key components of an effective performance measurement system include these:

- Clearly defined, actionable, and measurable goals that cascade from organizational mission to management and program levels;
- Cascading performance measures that can be used to measure how well mission, management, and program goals are being met;
- Established baselines from which progress toward the attainment of goals can be measured;
- Accurate, repeatable, and verifiable data; and
- Feedback systems to support continuous improvement of an organization's processes, practices, and results (FFC, 2004).

Qualitative and quantitative performance measures are being integrated into existing DOE project management practices and procedures (DOE, 2000). They are used at critical decision points and in internal and external reviews to determine if a project is ready to proceed to the next phase. Project directors and senior managers are using them to assess project progress and determine where additional effort or corrective actions are needed. However, DOE does not receive the full benefit of these measures because there is no benchmarking system to analyze the data to identify trends and successful techniques or compare actual performance with planned outcomes. For long-term process improvement, project performance measures and benchmarking processes should be used as projects are planned and executed as well as after they are completed.

Figure 2.1 describes a project performance control model that can be used to improve current and future projects by identifying trends and closing gaps between targeted and actual performance. Current DOE project and program management procedures such as Energy Systems Acquisition Advisory Board (ESAAB) reviews, Earned Value Management System (EVMS), Project Analysis and Reporting System (PARS), Office of Environmental Management Project Definition Rating Index (EM-PDRI), quarterly assessments, external independent reviews (EIRs), and independent project reviews (IPRs) are integrated into this model and called assessment processes.

In this model, project management processes are applied to inputs such as project resources to generate project plans, and these plans and resources become inputs for project execution. Individual projects are assessed and benchmarked against project targets and the performance of other projects. Output measures are compared with performance targets to identify performance gaps. These gaps are analyzed to identify corrective actions and improve the project as it proceeds. Once a project is completed, an assessment can be made of what worked well and where improvements in processes and project teams are needed for future projects (NRC, 2004c). The project outcomes are assessed to develop

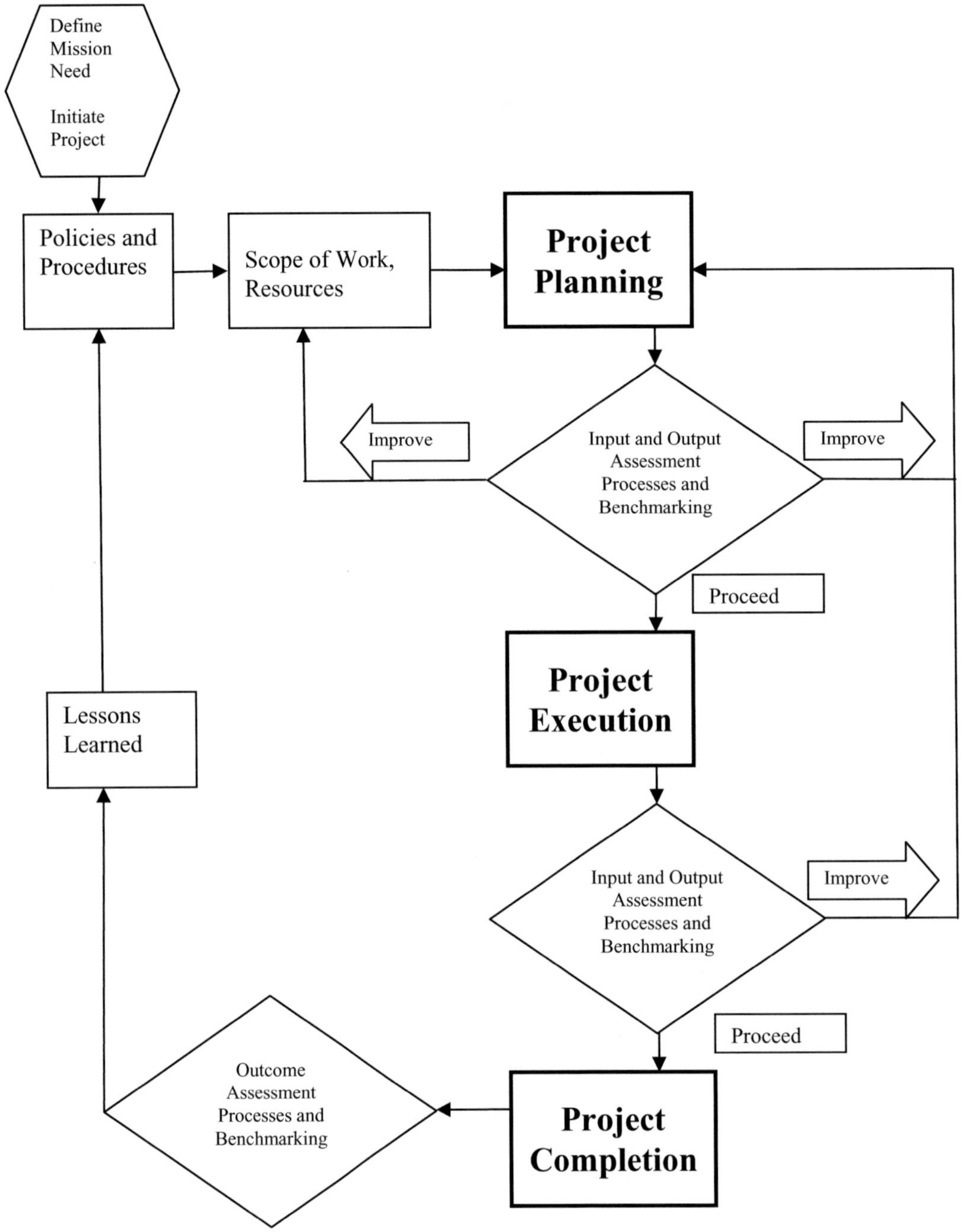

FIGURE 2.1 Model for controlling project performance.

lessons learned, which can be used as a feedback mechanism to improve policies and procedures and may drive changes in decision making and other processes.

INPUT, PROCESS, OUTPUT, AND OUTCOME MEASURES

Although assessment of the results of an internal process, such as project management, is much more straightforward than assessment of the results of public programs, the performance measures used can have intrinsic similarities. Performance measures for public program assessments are generally identified as input, process, output, and outcome (Hatry, 1999). Input is a measure of the resources (money, people, and time) provided for the activity being assessed. Process measures assess activities by comparing what is done with what should be done according to standard procedures or the number of process cycles in a period of time. Output measures assess the quantity and quality of the end product, and outcome measures assess the degree to which the end product achieves the program or project objectives. Assessment becomes more difficult as the target moves from input to outcome because of the influence of factors that are external to the program.

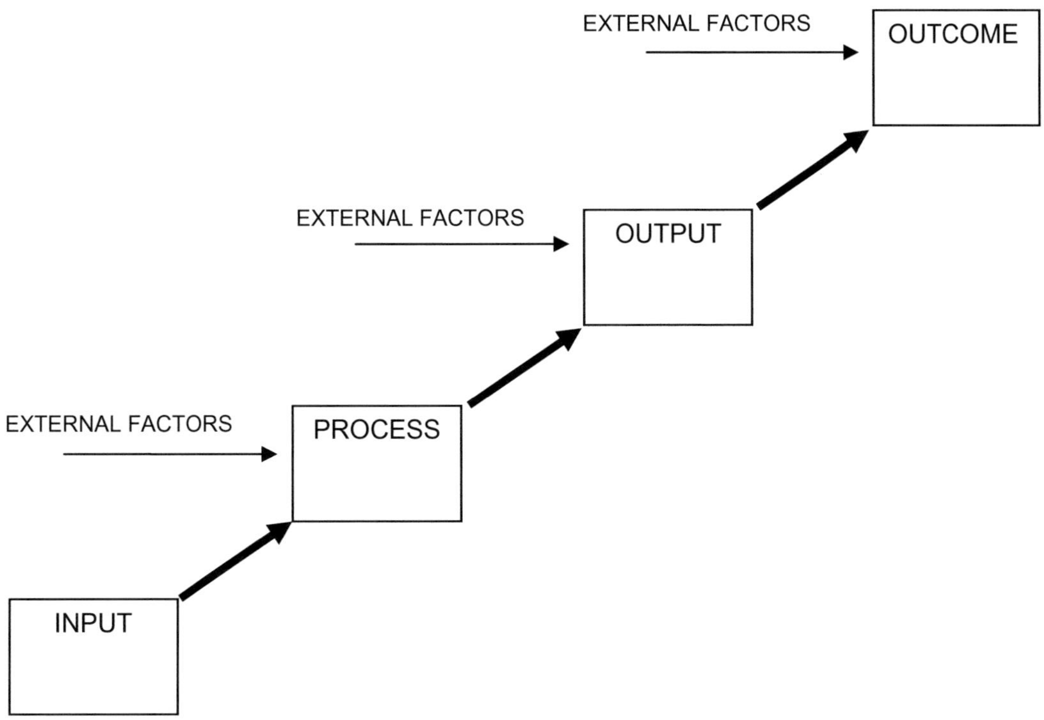

FIGURE 2.2 Performance assessment model.

Following this paradigm, project management is essentially a process; however, project management can be evaluated at both the program and the project level to assess its inputs, processes, outputs, and outcomes (Figure 2.2). At the program level, the input measures include the number of project directors and their training and qualifications. Program process measures relate to policies and procedures and how well they are followed. Program output measures identify how well projects are meeting objectives for cost and schedule performance. Outcome measures focus on how well the final projects support the program's or department's mission.

When project management is assessed at the project level, the input measures include the resources available and the quality of project management plans. Project process measures look at how well the plans are executed. Project output measures include cost and schedule variables, while outcome measures include scope, budget, and schedule and safety performance. In the 2003 assessment report (NRC, 2004a), the desired outcome at the program or departmental level was referred to as "doing the right project." The desired outcome at the project level was "doing it right." The committee noted that both are required for success.

The committee has identified all four types of measures, combined the two approaches (program and project), and grouped the measures in terms of the primary users (program managers and project managers). This approach separates measures used at the project level from measures used at the program/departmental level and combines input and process measures and output and outcome measures. In this approach, some output/outcome measures at the project level can be aggregated to provide measures at the program/departmental level through an analysis of distributions and trends. The committee believes that this will facilitate the benchmarking process by addressing the needs of the people who provide the data and derive benefits from the process.

SELECTING EFFECTIVE PERFORMANCE MEASURES

The measurement of performance is a tool for both effective management and process improvement. The selection of the right measures depends on a number of factors, including who will use them and what decision they support. For example, the airline industry has used on-time arrivals and lost bags per 1,000 as output measures, but to improve efficiency, procedures and processes are measured and analyzed in more detail by the airport ramp manager. Measures such as the time from arrival and chock in place to cargo door opening, the number of employees required and present for the type of aircraft, and whether the runner at the bottom of the conveyer is in place when the door is opened provide the information needed to improve efficiency and effectiveness.

Over the last several years DOE has improved and implemented new project management procedures and processes in the form of Order 413.3 (DOE, 2000). Efforts to measure and analyze implementation of the order at the project level can drive compliance and provide information for continued improvement. The committee believes that the requirements of Order 413.3 are directly correlated with project success but that performance data that measure the implementation of management plans are needed to support policies and guide future improvements.

The committee provides a set of performance measures in Tables 2.1 through 2.4, but it remains the responsibility of DOE to select and use the measures that work best for project directors, program offices, and the department. DOE should adopt performance measures suggested by the committee or other measures that have the following characteristics (NYSOT, 2003):

- Measurable, objectively or subjectively;
- Reliable and consistent;
- Simple, unambiguous, and understandable;
- Verifiable;
- Timely;
- Minimally affected by external influence;
- Cost-effective;
- Meaningful to users;
- Relate to mission outcome; and
- Drive effective decisions and process improvement.

The effectiveness of performance measures is also influenced by how well they are integrated into a benchmarking system. The system needs to be both horizontally and vertically integrated. That is, the measures need to be balanced to provide a complete assessment of the management of a project and be combinable across projects to assess the performance of the program and across programs to assess the impact of department-level policies and procedures.[1] If any organizational entity can identify a measure that has meaning and identity throughout an organization, such a measure is very valuable and should be the goal of performance measure development.

The committee's suggested performance measures are presented in four sets, including project-level input/process (Table 2.1), project-level output/outcome (Table 2.2), program- and department-level input/process (Table 2.3), and program- and department-level output/outcome (Table 2.4). Program- and department-level output/outcome are also the roll-up data of the project-level output/outcome measures. The program-level measures are also intended to be used throughout DOE. The tables describe the measures, how the measures are calculated, and specify the frequency with which data should be collected or updated.

Tables 2.1 through 2.4 include 30 performance measures. Taken individually, these measures lack robustness, but when they are analyzed as a group they provide a robust assessment of the individual variability and dependency of the performance measures. The adequacy of each performance measure individually is also limited, but combined they provide an assessment of the overall quality of project management for individual projects as well as the overall success of programs and the department. If the metrics are applied consistently over time and used for internal and external benchmarking, as described in Chapter 3, they will provide the information needed for day-to-day management and long-term process improvement.

REFERENCES

DOE (U.S. Department of Energy). 2000. Program and Project Management for Acquisition of Capital Assets (Order O 413.3). Washington, D.C.: U.S. Department of Energy.

EOP (Executive Office of the President). 1998. *A Report on the Performance-Based Service Contracting Pilot Project*. Washington, D.C.: Executive Office of the President.

FFC (Federal Facilities Council). 2004. *Key Performance Indicators for Federal Facilities Portfolios*. Washington, D.C.: The National Academies Press.

Hatry, H. 1999. *Performance Measurement: Getting Results*. Washington, D.C.: Urban Institute Press.

NRC (National Research Council). 2004a. *Progress in Improving Project Management at the Department of Energy, 2003 Assessment*. Washington, D.C.: The National Academies Press.

NRC. 2004b. *Intelligent Sustainment and Renewal of Department of Energy Facilities and Infrastructure*. Washington, D.C.: The National Academies Press.

NRC. 2004c. *Investments in Federal Facilities: Asset Management Strategies for the 21st Century*. Washington, D.C.: The National Academies Press.

NYSOT (New York State Office for Technology). 2003. *The New York State Project Management Guide Book, Release 2*. Chapter 4, Performance Measures. Available online at http://www.oft.state.ny.us/pmmp/guidebook2/. Accessed March 14, 2005.

Trimble, Dave. 2001. *How to Measure Success: Uncovering the Secrets of Effective Metrics.* Loveland, Col.: Quality Leadership Center, Inc. Available online at http://www.prosci.com/metrics.htm. Accessed April 25, 2005.

---

[1]Effective performance measures have also been referred to as SMART measures, meaning they are Specific, Measurable, Actionable, Relevant, and Timely (Trimble, 2001).

TABLE 2.1 Project Level Input/Process Measures

Project-Level Input/Process Measure	Comments	Calculation	Frequency
1. Implementation of project execution plan	Assumes appropriate plan has been developed and approved per O 413.3.	Assessment scale from 1 (poor) to 5 (excellent).[a] Poor: few elements implemented. Excellent: all elements required to date implemented.	Monthly
2. Implementation of project management plan	Assumes appropriate plan has been developed and approved per O 413.3.		
3. Implementation of project risk management plan	Assumes appropriate plan has been developed and approved per O 413.3.	Assessment scale from 1 (poor) to 5 (excellent). Poor: plan not reviewed since approved or few elements implemented. Excellent: plan reviewed and updated monthly and all elements required to date implemented.	Monthly
4. Project management staffing	Is project staffing adequate in terms of number and qualifications?	Assessment scale from 1 (poor) to 5 (excellent).	Quarterly
5. Reliability of cost and schedule estimates	This measure is collected for each baseline change, and reports include current and all previous changes.	Number of baseline changes. For each baseline change, (most recent baseline less previous baseline) divided by previous baseline.	Quarterly
6. Accuracy and stability of scope	Does the current scope match the scope as defined in CD-0?	Assessment of the impact of scope change on project costs and schedule on a scale from 1 (substantial) to 5 (none). Cost impacts of scope changes since CD-0 are 1. Substantial, >10% 2. Significant, 5-10% 3. Moderate, 2.5-5% 4. Minor, <2.5% 5. None. Schedule impacts of scope changes since CD-0 are 1. Substantial, >10% 2. Significant, 5-10% 3. Moderate, 2.5-5% 4. Minor, <2.5% 5. None. Scope currently adheres to the scope defined at CD-0.	Quarterly

TABLE 2.1 Continued

Project-Level Input/Process Measure	Comments	Calculation	Frequency
7. Effectiveness of project communication	Adapt existing criteria and measurements from existing tools such as those developed by the Construction Industry Institute.	Assessment of seven key factors of project communication on a scale from 1 (poor) to 5 (excellent). Overall effectiveness of project communications. Accuracy of information. Timeliness of information. Completeness of information. Understanding of information. Barriers to communication from 1 (significant) to 5 (none). Communication of procedures. Assessment of overall project communication is the sum of the evaluation of individual factors 7 (poor) to 35 (excellent).	Semiannually
8. Corrective actions	Corrective action from IPRs, EIRs, and quality assurance reports.	Percent correction actions open.	Monthly

[a]Applies to both project execution plan and project management plan.

TABLE 2.2 Project-Level Output/Outcome Measures

Project-Level Output/Outcome Measure	Comments	Calculation	Frequency
1. Cost growth		(Estimated cost at completion less CD-2 baseline cost) divided by CD-2 baseline cost.	Monthly
		(Actual cost at completion less CD-2 baseline cost) divided by CD-2 baseline cost.	End of project
2. Schedule growth		(Estimated duration at completion less CD-2 baseline duration) divided by CD-2 baseline duration.	Monthly
		(Actual duration at completion less CD-2 baseline duration) divided by CD-2 baseline duration.	End of project
3. Phase cost factors	Use CD phases	Actual project phase cost divided by actual cost at completion.	End of project
4. Phase schedule factors	Use CD phases	Actual project phase duration divided by total project duration.	End of project
5. Preliminary engineering and design (PE&D) factor		PE&D funds divided by final total estimated cost.	End of project
6. Cost variance	As currently defined by EVMS	Budgeted cost of work performed less actual cost of work performed.	Monthly
7. Cost performance index	As currently defined by EVMS	Budgeted cost of work performed divided by actual cost of work performed.	Monthly
8. Schedule variance	As currently defined by EVMS	Budgeted cost of work performed less budget cost of work scheduled.	Monthly
9. Schedule performance index	As currently defined by EVMS	Budgeted cost of work performed divided by budgeted cost of work scheduled.	Monthly
10. Safety performance measures	As defined by DOE environment safety and health (ES&H) policies	Total recordable incident rate (TRIR) and days away restricted and transferred (DART).	Monthly

TABLE 2.3 Program- and Department-Level Input/Process Measures

Program- and Department-Level Input/Process Measure	Comments	Calculation	Frequency
1. Project director staffing.	The number of certified project directors offers a shorthand way to assess the adequacy of senior-level staffing for major projects. In addition, information relating to their certification levels and experience can provide useful indicators over time of the health of the project management development process.	Ratio of project directors to the number of projects over $5 million. Ratio of project directors to the total dollar value of projects. Number and percentage of projects where project directors have the appropriate level of certification. Average years of experience of the project director community.	Quarterly
2. Project support staffing.	By identifying how many federal staff are assigned to a project and then over time measuring the project's cost and schedule performance, DOE may develop some useful rules of thumb on optimal levels of staffing. To ensure a consistent count, staff should be formally assigned to the project and spend at least 80% of their time working on it.	Ratio of federal staff to the numbers of projects over $5 million. Ratio of federal staff to the total dollar value of projects. Mix of skill sets found in the federal project staff.	Quarterly
3. Senior management involvement.	The commitment of DOE leadership to good project management can be an important factor in ensuring continuing progress. This commitment can most easily be seen in the time that leadership devotes to project management issues and events. Senior leadership's presence at critical decision and quarterly reviews sends a strong signal of the importance DOE attaches to good project performance.	Frequency of critical decision reviews rescheduled. Effectiveness of reviews. Frequency of substitutes for senior leadership scheduled to conduct the reviews. Assessment scale from 1 (poor) to 5 (excellent): 1. Senior management absent from more than half of critical decision and quarterly reviews and reviews are frequently rescheduled. 2. Senior management present at and engaged in 50-75% of reviews; some reviews rescheduled. 3. Senior management present at and engaged in 75-90% of reviews; some reviews rescheduled.	Annually

Continues

TABLE 2.3 Continued

Program- and Department-Level Input/Process Measure	Comments	Calculation	Frequency
		4. Senior management present at and engaged in >90% of reviews; all reviews as scheduled. 5. Senior management present and engaged in all reviews; all reviews as scheduled.	
4. Commitment to project management training.	Ensuring that key project management personnel are adequately prepared to carry out their assignments is critical to project management success. The willingness of DOE to commit resources to ensure that staff are able to fulfill their roles is an important indicator of the department's support for improving project performance.	Adequacy of data available on project staff training. Correlation over time between levels of project staff training and project success. Comparison of the level and amount of training for DOE federal project staff with that for the staff of other departments (e.g., Defense) and the private sector.	Annually
5. Identification and use of lessons learned.	By effectively using lessons learned from other projects, the federal project director can avoid the pitfalls experienced by others.	Categories for defining good project management from lessons-learned data. Comparison of factors used by DOE with factors used by the private sector to indicate effective project management performance. Suggested assessment from 1 (ineffective) to 5 (highly effective): 1. No attempt to collect lessons-learned information. 2. Lessons learned collected but on a haphazard basis. 3. Systematic approach established to collect lessons-learned information. 4. Lessons-learned information collected on a regular basis and available for review. 5. Lessons-learned data collected, stored, and acted on regularly.	Annually

TABLE 2.3 Continued

Program- and Department-Level Input/Process Measure	Comments	Calculation	Frequency
6. Use of independent project reviews (external and internal reviews) and implementation of corrective action plans.	Independent reviews can alert project directors to potential problem areas that will need to be resolved as well as provide suggestions for handling these problems. The varied experience of the independent reviewers offers the project director broad insight that goes beyond that available through his or her project team.	Was an external independent review conducted on the project prior to CD-3? Percent of EIR issues addressed by corrective actions.	Quarterly
7. Number and dollar value of performance-based contracts.	Performance-based contracting methods improve contractors' focus on agency mission results. An increase in the number and value of such contracts should lead to improved project performance. A 1998 study conducted by the Office of Federal Procurement Policy found cost savings on the order of 15% and increases in customer satisfaction (up almost 20%) when agencies used performance-based contracts rather than traditional requirements-based procurements (EOP, 1998). Much of the success of a performance-based contracting approach results from effective use of a cross-functional team for identifying desired outcomes and establishing effective performance metrics.[a]	Number performance-based service contracts. Total dollar value of performance-based contracts. Distribution of performance-based contracts across the various DOE programs. Share of performance-based contracts in all service contracts let by the department (dollar value). Percentage of project management staff who have received performance-based contracting training. Number of contracts that employed performance incentives to guide contractor actions. Number of contracts that were fixed price and the number that were cost plus.	Annually
8. Project performance as a factor of project size and contract type. Number and dollar value of contracts at the following funding levels: Under $5 million $5 million to $20 million $20 million to $100 million Over $100 million	Information on the level and size of contracting actions will assist DOE in determining whether there is significant variance in project management performance depending on the overall dollar value of the project. To make this determination, data are needed on project cost and schedule performance for each project against initial targets.	Variance in project cost and schedule performance based on the funding level of the project. Variance in project performance across the programs of the department. Correlation between use of various contracting techniques and project success.	Annually

Continues

TABLE 2.3 Continued

Program- and Department-Level Input/Process Measure	Comments	Calculation	Frequency
9. Use of good project management practices.	Following good project management practices should lead to better project performance. Development, implementation, and timely updating of a good acquisition plan and an effective risk mitigation strategy are examples of such practices. Others include the following: • Use of teamwork, • Regularly assessing performance and reviewing and analyzing lessons learned at project completion, • Benchmarking the project against others' successful efforts, and • Ensuring that all project staff see how the project furthers the department's mission	Percent of project staff trained in the application of O413.3. Are procedures in place to evaluate on a regular basis how well project staff carry out each of the elements of O413.3? Suggested approach for assessing the effectiveness of project management performance, from 1 (ineffective) to 5 (highly effective): 1. No attempt to follow guidelines in O 413.3. 2. Partial documentation prepared but not followed. 3. Required documentation prepared but not followed or updated. 4. Required documentation prepared but performance assessment sporadic and updates incomplete or not timely. 5. Practices documented, followed, updated on a timely basis; performance assessed and corrective actions taken as needed.	Annually

[a]In a September 7, 2004, memorandum for chief acquisition officers and senior procurement executives, Robert A. Burton, associate administrator of the Office of Federal Procurement Policy, Office of Management and Budget, stated that "agencies should apply PBSA [performance-based systems acquisition] methods on 40 percent of eligible service actions over $25,000, to include contracts, task orders, modifications and options, awarded in fiscal year (FY) 2005, as measured in dollars."

TABLE 2.4 Program- and Department-Level Output/Outcome Measures

Program- and Department-Level Output/Outcome Measure	Comments	Calculation	Frequency
1. Assessment of project performance.	DOE uses a dashboard-type approach for identifying to department executives those projects that are within budget and on schedule and those that are experiencing various levels of difficulty. An accounting of the number and percent of projects identified as green (variance within 0.90 to 1.15 of baseline cost and schedule), yellow (variance within 0.85 to 0.90 or 1.15 to 1.25 of baseline cost and schedule), or red (baseline variance of cost and schedule <0.85 or >1.25) offers a quick and simple means for assessing overall project performance across the programs over time. Progress is achieved as fewer projects are found with the yellow and red designations.[a]	Number and percentages of projects designated as green, yellow, and red. Change in percentages of projects designated as green, yellow, and red over time.	Quarterly
2. Assessment of the length of time projects remain designated yellow or red.	This measures the effectiveness of senior management review and the ability of project staff to take corrective action.	Average length of time a project is designated yellow. Average length of time a project is designated red. Change in average length of time a project is designated yellow or red for programs over time.	Quarterly
3. Mission effectiveness.	This measure is intended to integrate project performance measures with facility performance measures.[b] It is a comparative measure of an individual project within the program and how effectively the completed project fulfills its intended purposes and if the program has undertaken the "right project."	Assessment scale from 1 (poor) to 5 (excellent): 1. Facility does not support program mission. 2. Facility does not support intended mission but can be used for another mission. 3. Facility supports mission with major modification. 4. Facility supports mission with minor modification. 5. Facility supports program mission.	End of project

[a]DOE's current dashboard assessment prepared for senior management is basically effective. However, it depends on data in PARS that are not currently verified, and there is no process for collecting and analyzing historical data. Also, the committee believes the effectiveness of the dashboard in assessing project performance would be improved if the values for cost and schedule were differentiated and the expected performance were raised to green (0.95 to 1.05), yellow (0.90 to 0.95 or 1.05 to 1.10), and red (<0.90 or >1.10) for cost variance and to green (0.95 to 1.10), yellow (0.90 to 0.95 or 1.10 to 1.20), and red (<0.90 or >1.20) for schedule variance.

[b]See *Intelligent Sustainment and Renewal of Department of Energy Facilities and Infrastructure,* "Integrated Management Approach" (NRC, 2004b, p. 62).

3
The Benchmarking Process

INTRODUCTION

Management theory and practice have long established a link between effective performance measures and effective management (Drucker, 1995). The effectiveness of any given performance measure depends on how it will be used. For performance measures to have meaning and provide useful information, it is necessary to make comparisons. The comparisons may evaluate progress in achieving given goals or targets, assess trends in performance over time, or weigh the performance of one organization against another (Poister, 2003).

The Government Performance and Results Act of 1993 (GPRA) established the requirement for performance measures to assess how well departments and agencies are achieving their stated goals and objectives. The emphasis of GPRA performance measures is on output and outcome measures at the program level.

Performance measures used as a management tool need to be broadened to include input and process measures. One approach is to use an array or scorecard composed of multiple measures. The Balanced Scorecard is one such approach that assesses an organization and its programs from four different perspectives: customer, employee, process, and finance. "The scorecard creates a holistic model of the strategy that allows all employees to see how they contribute to organizational success . . . . [It] focuses change efforts. If the right objectives and measures are identified, successful implementation will likely occur." (Kaplan and Norton, 1996, p. 148)

The objectives and process for construction and construction project management create a good environment for the effective use of benchmarking for measuring and improving performance. Benchmarking is a core component of continuous improvement programs. As Gregory Watson noted in his *Benchmarking Workbook*, 12 of the 32 criteria for the Malcolm Baldrige National Quality Award refer to benchmarking as a key component of quality assurance and process improvement (Watson, 1992). The role of benchmarking in process improvement is similar to that of the Six Sigma[1] process improvement methodology. The Six Sigma methodology comprises five integrated steps: define, measure, analyze, improve, and control (DMAIC). These steps are also central to the benchmarking process defined in this chapter.

Benchmarking is an integral part of the continuous improvement cycle shown in Figure 3.1 (CII, 2004). Measuring, comparing to competition, and identifying opportunities for improvements are the essence of benchmarking.

---

[1]Six Sigma refers to a body of statistical- and process-based (e.g., process mapping, value stream mapping) methodologies and techniques used as part of a structured approach for solving production and business process problems plagued with variability in execution (Harry and Schroeder, 2000).

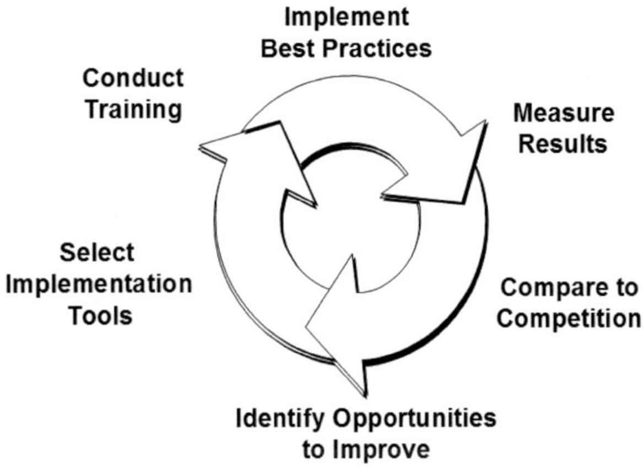

FIGURE 3.1 Continuous improvement cycle. SOURCE: CII, 2004.

BENCHMARKING ROADMAP

Many definitions of benchmarking are available. The following definition, from the Construction Industry Institute (CII), illustrates a number of important points.

> Benchmarking is the systematic process of measuring one's performance against recognized leaders for the purpose of determining best practices that lead to superior performance when adapted and utilized. (CII, 1995)

To be successful, benchmarking should be implemented as a structured, systematic process. It will not be successful if applied in an ad hoc fashion on a random basis. In most cases benchmarking is best-practice-oriented and is part of a continuous improvement program that incorporates a feedback process. Benchmarking requires an understanding of what is important to the organization (sometimes called critical success factors) and then measuring performance for these factors. The gap between actual performance and preferred achievement is typically analyzed to identify opportunities for improvement. Root cause analysis usually follows to assess the cause of unsatisfactory performance, and a search for best practices may be used to help address performance problems. Figure 3.2 illustrates the process with a benchmarking roadmap.

The roadmap was adapted from the 10-step process introduced by Robert Camp at Xerox. Camp pioneered much work in benchmarking, and some even credit him with the first use of the term "benchmarking."

EXTERNAL VERSUS INTERNAL BENCHMARKING

Benchmarking can be internal or external. When benchmarking internally, organizations benchmark against their own projects. When benchmarking externally, organizations seek projects from other companies or perhaps, in the case of DOE, from separate program offices for comparative analysis. External benchmarks are generally considered to provide the greater advantage; however, internal benchmarking can be useful where no external benchmarks are available. Internal benchmarks are often the starting point for quantitative process examination. Trends can be identified by examining these data over time, and the impact of performance-improving processes can be assessed. External benchmarks provide the added advantage of comparing against competitors. Without external benchmarks, an organization and its managers may lack an understanding of what constitutes "good" performance.

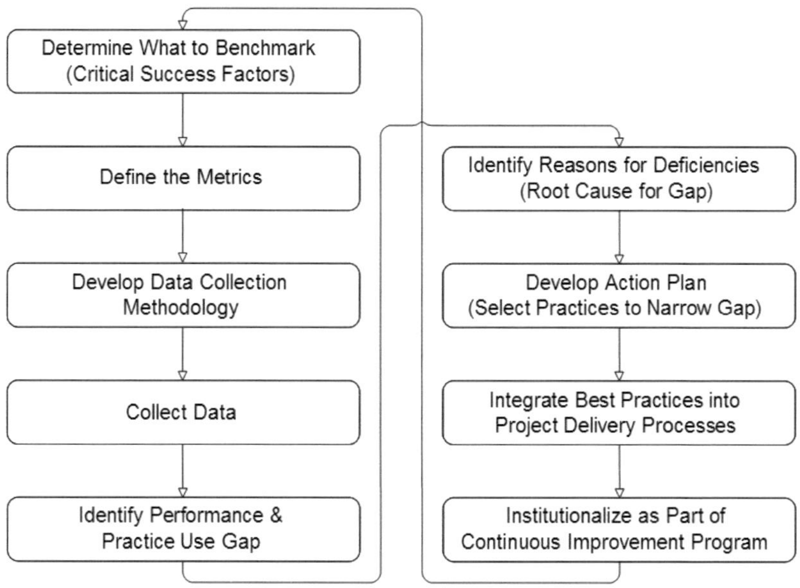

FIGURE 3.2 Benchmarking roadmap. SOURCE: Adapted from Camp, 1989.

Application at Various Project Phases and Management Levels

Benchmarking can and should be used at various levels throughout the organization, but if project improvement is the goal, data will typically be entered at the project level. Program- and department-level measures can be provided by roll-ups of the project-level data.

Benchmarking can be applied during various phases of a project for different purposes. When applied early on, such as at project authorization, it can be used to identify characteristics that may be associated with potential future problems and to identify aspects of project management (e.g., risk management) that need special attention to ensure project success. When applied during project execution, it can serve as a project management tool to guide project decisions. Postproject benchmarking is usually used to assess performance of a project delivery system to provide for lessons learned and feedback that can be used to establish benchmarks for future comparisons. Most organizations tend to begin with postproject comparisons and later progress to the earlier uses as confidence in the benchmarking process builds. Over time, when sufficient data are available, trends can be analyzed to provide insight into the performance of project management systems. Since integrated project team (IPT) members will normally have moved on to new projects, trend analyses of project-level cost and schedule metrics would typically be used at program and department levels.

Benchmarking needs buy-in at various levels of an organization in order to be successful. Most often, benchmarking is driven from the top. Senior management commitment is critical if resources are to be made available for the process. While benchmarking may succeed with senior management support alone, it is far more likely to succeed if it has the support of middle management and the project team. Furthermore, the project team is far more likely to support the benchmarking initiative if it is understood that the goal is system improvement and not individual or team performance appraisal. The IPT members should be confident that data submitted for benchmarking will not be used for performance appraisals if accurate data are to be obtained.

Validation

The validation of benchmarked data is a critical component of any benchmarking system. Some benchmarking services collect data through a survey instrument and then use an experienced analyst to review them. The project team is interviewed to clarify and resolve issues.

A different approach to validation is to share responsibility between the project team and an outside organization. The project team is responsible for reviewing the data to be submitted to ensure that they accurately reflect the project's experience. An independent reviewer serves as an honest broker and validates the data by ensuring their completeness and accuracy. The reviewer should be a trained, independent professional with a good understanding of the data to be collected, the measures to be produced, and the project management process used. A rigorous examination of all data is performed by the service provider as a final check. Whatever approach is used, a validation process assists in maintaining consistency across organizations.

IMPLEMENTATION OF BENCHMARKING

Benchmarking processes are not easy to implement, and to be successful an organization must overcome numerous barriers. Some private-sector companies fear that they may lose their competitive advantage by sharing information, and others fear exposure of organizational weakness. Use of an identity-blind process, whereby data are posted without attribution, can mitigate these concerns.

For some organizations, arrogance is a major obstacle. These organizations may believe they are the best, so why benchmark? As renowned management consultant W. Edwards Deming would probably ask superconfident organizations that lack performance data and comparison to other organizations: How do you know? (Watson, 1992). Other organizations are unaware of the value of benchmarking and believe that benchmarking systems do not adequately address their needs. Benchmarking agreements and training increase familiarity with the benchmarking process and can help to reduce these barriers.

One of the greatest barriers to benchmarking is a lack of resources. Most organizations are leaner today than in the past, and dedicating the essential resources can be difficult. For some organizations, project processes and computer systems are not sufficiently developed to easily support benchmarking (CII, 2002). For these organizations the benchmarking process will require more manual intervention and consequently greater resources. As project processes become automated, this barrier should shrink.

Lessons Learned

Lessons learned from past benchmarking efforts can be helpful for an organization embarking on a benchmarking initiative:

- Senior management buy-in and support are vital to success, but even with this support, generating enthusiasm is difficult (McCabe, 2001).
- Department- and program-level champions are essential.
- Even though projects may be unique, the processes are very similar.
- A high code of ethics is essential.
- Benchmarking will be successful only if made an integral part of the project process.
- Commonly accepted, effective metrics for assessing project performance are necessary to assess the extent to which best practices are used. Input, process, output, and outcome performance measures are necessary, and it is possible to implement them.
- Performance measures should be applied through a structured benchmarking process.
- Cost-effective, value-added benchmarking can be implemented through standardization of definitions and application of computer-based technologies.

REFERENCES

Camp, Robert C. 1989. *Benchmarking, The Search for Industry Best Practices That Lead to Superior Performance.* Portland, Ore.: ASQC Quality Press.

CII (Construction Industry Institute). 1995. *Construction Industry Institute Data Report.* Austin, Tex.: Construction Industry Institute.

CII. 2002. Member Company Survey. Austin, Tex.: Construction Industry Institute.

CII. 2004. Presentation to the Construction Industry Institute Annual Conference, Vancouver, British Columbia, July 2004. Austin, Tex.: Construction Industry Institute.

Drucker, Peter F. 1995. *Managing in a Time of Great Change.* New York, N.Y.: Penguin Putnam, Inc.

Kaplan, Robert S., and David P. Norton. 1996. *The Balanced Scorecard.* Boston, Mass.: Harvard Business School Press.

McCabe, Steven. 2001. *Benchmarking in Construction.* Oxford, U.K.: Blackwell Science, Ltd.

Mikel, Harry, and Richard Schroeder. 1996. *SIX SIGMA: The Breakthrough Management Strategy Revolutionizing the World's Top Corporations.* New York, N.Y.: Currency/Doubleday.

Poister, Theodore H. 2003. *Measuring Performance in Public and Nonprofit Organizations.* San Francisco, Calif.: Jossey-Bass.

Watson, Gregory H. 1992. *The Benchmarking Workbook: Adapting Best Practices for Performance Improvement.* Portland, Ore.: Productivity Press.

4
Implementation

INTRODUCTION

The committee considers effective performance measurement and a benchmarking process to be essential activities of successful project management organizations. The committee notes that benchmarking is one of many tools that can be used for process improvement. In the case of project management, benchmarking is particularly useful because external benchmarks have already been developed by the Construction Industry Institute (CII) and others. Such external benchmarks can help develop an understanding of what constitutes good performance. As noted in Chapter 1, the NRC's 2003 assessment report concluded that shortcomings in these areas persist in DOE (NRC, 2004a). In the present report, the committee provides guidance for establishing a system that will enable DOE to measure project management performance at both the project and program levels and to benchmark internally within and between program secretarial offices (PSOs) and externally with other federal agencies and the private sector. The corrective actions required for continued improvement of project management can be based on information gained in this way. The information provided in this report should be viewed not as the final word but rather as a first step toward development of a viable methodology to suit the needs and goals of DOE.

The establishment of an effective and sustainable performance measurement and benchmarking system requires the commitment of top management and of resources. While maintaining such a system is not without cost, its value has been demonstrated within the government and private-sector organizations (NPR, 1997).

This report does not contain new findings and recommendations. It is intended to provide guidance for implementing previous NRC recommendations on performance measures and benchmarking. Further, the committee believes specific recommendations on the implementation of performance measures and benchmarking would be inappropriate. For a system to be effective it should be crafted to fill multiple organizational needs and it should carry the imprimatur of the users and be accepted at all levels of the organization.

Measuring performance and benchmarking should be viewed as a routine, integral part of the project management process rather than a separate function. However, integration into the personnel appraisal system is to be avoided. Accountability for project outcomes continues to be important; however, benchmarking should emphasize the continuous and consistent application of the system and the resultant identification and implementation of improved project management procedures through lessons learned and feedback systems.

DEVELOPMENT OF A PERFORMANCE MEASUREMENT AND BENCHMARKING SYSTEM

The performance measures presented in Chapter 2 are fundamental and should be considered as a starting point for the development of a performance measurement and benchmarking system for DOE.

While additions and adjustments will be necessary to accommodate particular situations, consistency is essential for meaningful benchmarking. To assure department-wide consistency, it will be necessary to agree on the definition of the key performance measures, how they will be calculated, and a reporting format. Senior management will need to show its commitment and champion the process at all levels of the department. OECM should lead the process, but program offices and project directors and managers in the field need to play an active role as well.

Gathering data for calculating the various performance measures will largely be done by individuals at the project level. For this to happen with an acceptable level of certainty, the responsible individuals should be inculcated with the need for accuracy and the benefits of benchmarking and be committed to implementing the system. Above all, it must be made clear that no penalties or recriminations will be incurred for honest reporting. A high code of ethics must prevail. Although an outside contractor could help to establish the system and gather data in the early stages, the working input to the system must come from DOE personnel.

This report largely addresses the first four steps of the benchmarking roadmap presented in Figure 3.2:

- Determine what to benchmark,
- Define the measures,
- Develop data collection methodology, and
- Collect data.

The remaining five steps define the essence and purpose of benchmarking—that is, to continuously improve project management. For any benefits to be derived from the effort, it is essential that DOE establish a means for implementing the steps beyond the collection of data. This will require advocacy to be built into the system. The committee believes that corrective improvements to the project management process should be guided from within DOE, not from outside. As noted in the 2003 assessment report, this process should be led by a senior management champion with the full participation of OECM and the PSOs (NRC, 2004a).

Benchmarking data should not be used for individual performance reviews, but a project director's application of the benchmarking process should be assessed. Performance measurement and benchmarking need to be part of DOE's Project Manager Career Development Program beginning at the first level of certification. This will require specific training in the application of DOE-specific benchmarking procedures and performance measures. The training should be detailed enough to enable project directors to benchmark their projects as well as contribute to the development and improvement of the measures and process. The investment in training will also communicate the importance of benchmarking.

DATA COLLECTION AND VALIDATION

The committee believes that one of the first ways to make benchmarking meaningful and consistent is to ensure that the Earned Value Management System (EVMS) is being used consistently throughout the department. The DOE Project Analysis and Reporting System (PARS) should be expanded to become an integral part of the benchmarking data collection system. In order to serve this function PARS needs to be a robust, user-friendly, real-time, electronic system that accommodates the necessary data elements and produces the comparisons in tabular and graphic form. While the extant PARS may serve as a base for the benchmarking, it will need to be expanded before it can become a useful tool for benchmarking at all levels.

Any benchmarking system requires periodic evaluation to ensure that the basic definitions of the measures are understood, that the data being reported are consistent and accurate, and that the information is being used to make corrections and improve performance. The generation and input of data as well as the validation of these data are inherently a function of the owner, in this case DOE. Owners may engage contractors to assist in the collection, validation, and analysis of the data, but the owner's management team needs to be involved throughout the process.

BENCHMARKING READINESS

For benchmarking to succeed, an organization needs to have the necessary technical, managerial, and cultural mechanisms in place (Keehly, 1997). A common set of performance measures needs to be developed that is consistent within the organization and understood by external benchmarking partners. The performance measures also need to be integrated into a common data collection system. As noted above, project directors need to be trained and a system for validating data needs to be established. The most difficult step is establishing an organizational culture that is ready to assess, compare, and analyze performance and to adopt best practices used by others when they are identified. This requires an organizational commitment to continuous improvement, acceptance of new ideas, and open communication and cooperation at all levels of the organization.

Organizational readiness for benchmarking can be developed incrementally, starting with a limited number of measures and internal benchmarking within a program, then expanding the effort to include more diverse measures and comparisons with other programs. The objective over time should be to develop a full set of measures and to benchmark both internally and externally, against organizations in other government agencies and private industry.

REFERENCES

Keehly, Patricia, S.Medlin, S. MacBride, and L. Longmire, 1997. *Benchmarking for Best Practices in the Public Sector*. San Francisco, Calif.: Jossey-Bass.

NPR (National Partnership for Reinventing Government Reports, formerly the National Performance Review). 1997. *Serving the American Public: Best Practices in Performance Measurement.* Available online at http://govinfo.library.unt.edu/npr/library/papers/benchmrk/nprbook.html. Accessed March 14, 2005.

NRC (National Research Council). 2004. *Progress in Improving Project Management at the Department of Energy, 2003 Assessment*. Washington, D.C.: The National Academies Press.

Appendixes

A
NRC's DOE Project Management Assessment Reports

1998 *Assessing the Need for Independent Project Reviews in the Department of Energy*

1999 *Improving Project Management in the Department of Energy*

2001 "Improving Project Management in the Department of Energy." *Letter report, January*

2001 *Progress in Improving Project Management at the Department of Energy: 2001 Assessment*

2002 "Progress in Improving Project Management at the Department of Energy: 2002 Interim Assessment." *Letter report, May*

2002 *Proceedings of Government/Industry Forum: The Owner's Role in Project Management and Preproject Planning*

2003 *Progress in Improving Project Management at the Department of Energy: 2002 Assessment*

2004 *Progress in Improving Project Management at the Department of Energy: 2003 Assessment*

2005 *The Owner's Role in Project Risk Management*

B
Statement of Task

In response to a congressional directive, House Report 106-253, this effort will review and comment on DOE's recent efforts to improve its project management, including (1) specific changes implemented by the department to achieve improvement (e.g., organization, practices, training), (2) an assessment of the progress made in achieving improvement, and (3) the likelihood that improvement will be permanent. This activity is the third in a series requested by Congress regarding the delivery and management of DOE projects. The language of the congressional directive requires the NRC to produce semiannual assessment progress reports.

This project was initially planned as a 3-year assessment from July 2000 through September 2003. The committee was extended at the sponsor's request to provide additional information and more detailed guidance for implementing recommendations on improving project risk management and benchmarking. The committee's supplemental report on the owner's role in project risk management was published in January 2005. This report on performance measures and benchmarking completes the committee's tasks.

C
Biographies of Committee Members

Lloyd A. Duscha (National Academy of Engineering), *Chair*, retired from the U.S. Army Corps of Engineers in 1990 as the highest-ranking civilian after serving as deputy director, Engineering and Construction Directorate, at headquarters. He has expertise in engineering and construction management, was a principal investigator for the NRC report *Assessing the Need for Independent Project Reviews in the Department of Energy,* and was a member of the committee that produced the NRC report *Improving Project Management in the Department of Energy*. He served in numerous progressive Army Corps of Engineer positions in various locations over four decades. Mr. Duscha is currently an engineering consultant to various national and foreign government agencies, the World Bank, and private sector clients. He has served on numerous NRC committees and recently served on the Committee on the Outsourcing of the Management of Planning, Design, and Construction Related Services as well as the Committee on Shore Installation Readiness and Management. He chaired the NRC Committee on Research Needs for Transuranic and Mixed Waste at Department of Energy Sites and serves on the Committee on Opportunities for Accelerating the Characterization and Treatment of Nuclear Waste. He has also served on the Board on Infrastructure and the Constructed Environment and was vice chairman for the U.S. National Committee on Tunneling Technology. Other positions held were president, U.S. Committee on Large Dams; chair, Committee on Dam Safety, International Commission on Large Dams; executive committee, Construction Industry Institute; and the board of directors, Research and Management Foundation of the American Consulting Engineers Council. He has numerous professional affiliations, including fellowships in the American Society of Civil Engineers and in the Society of American Military Engineers. He holds a B.S. in civil engineering from the University of Minnesota, which awarded him the Board of Regents Outstanding Achievement Award.

Don Jeffery (Jeff) Bostock recently retired from Lockheed Martin Energy Systems, Inc., as vice president for engineering and construction with responsibility for all engineering activities within the Oak Ridge nuclear complex. In addition to his experience with managing projects as a DOE contractor, he has also served as vice president of defense and manufacturing and as manager of the Oak Ridge Y-12 plant, a nuclear weapons fabrication and manufacturing facility. His career at Y-12 included engineering and managerial positions in all of the various manufacturing, assembly, security, and program management organizations. He also served as manager of the Paducah Gaseous Diffusion Plant, providing uranium enrichment services. He was a member of the committees that produced the NRC reports *Proliferation Concerns: Assessing U.S. Efforts to Help Contain Nuclear and Other Dangerous Materials, Technologies in the Former Soviet Union,* and *Protecting Nuclear Weapons Material in Russia*. Mr. Bostock also served as a panel member for the annual NRC assessment of the National Institute of Standards and Technology Measurement and Standards Laboratories. Mr. Bostock has a B.S. in industrial engineering from Pennsylvania State University and an M.S. in industrial management from the University of Tennessee. He is a graduate of the Pittsburgh Management Program for Executives.

Allan V. Burman is president of Jefferson Solutions, a division of the Jefferson Consulting Group, a firm that provides change management services and acquisition reform training to many federal departments and agencies. He has expertise in federal acquisition, procurement, and budget reform. Dr. Burman provides strategic consulting services to private sector firms doing business with the federal government as well as to federal agencies and other government entities. He also has advised firms, congressional committees, and federal and state agencies on a variety of management and acquisition reform matters. Prior to joining the Jefferson Group, Dr. Burman had a lengthy career in the federal government, including as administrator for federal procurement policy in the Office of Management and Budget (OMB), where he testified before Congress over 40 times on management, acquisition, and budget matters. Dr. Burman also authored the 1991 policy letter that established performance-based contracting and greater reliance, where appropriate, on fixed-price contracting as the favored approach for contract reform. As a member of the Senior Executive Service, Dr. Burman served as chief of the Air Force Branch in OMB's National Security Division and was the first OMB branch chief to receive a Presidential Rank Award. Dr. Burman is a fellow and member of the board of advisors of the National Contract Management Association, a principal of the Council for Excellence in Government, a director of the Procurement Round Table, and an honorary member of the National Defense Industrial Association. He was also a contributing editor and writer for *Government Executive* magazine. Dr. Burman obtained a B.A. from Wesleyan University, was a Fulbright scholar at the Institute of Political Studies, University of Bordeaux, France, and has a graduate degree from Harvard University and a Ph.D. from the George Washington University.

G. Brian Estes is the former director of construction projects, Westinghouse Hanford Company, where he directed project management functions for construction projects in support of operations and environmental cleanup at the Department of Energy's Hanford nuclear complex. His has experience with DOE, as well as other large-scale government construction and environmental restoration projects. He served on the committee that produced the recent NRC report *Improving Project Management in the Department of Energy* and on a number of other NRC committees. Prior to joining Westinghouse, he completed 30 years in the Navy Civil Engineer Corps, achieving the rank of rear admiral. Admiral Estes served as commander of the Pacific Division of the Naval Facilities Engineering Command and as commander of the Third Naval Construction Brigade, Pearl Harbor, Hawaii. He supervised over 700 engineers, 8,000 Seabees, and 4,000 other employees in providing public works management, environmental support, family housing support, and facility planning, design, and construction services. As vice commander, Naval Facilities Engineering Command, Admiral Estes led the total quality management transformation at headquarters and two updates of the corporate strategic plan. Admiral Estes directed execution of the $2 billion military construction program and the $3 billion facilities management program while serving as deputy commander for facilities acquisition and deputy commander for public works, Naval Facilities Engineering Command. He holds a B.S. in civil engineering from the University of Maine and an M.S. in civil engineering from the University of Illinois and is a registered professional engineer in Illinois and Virginia.

Steven L. Fahrenkrog is director of the Project Management Institute's (PMI's) Knowledge Delivery Group, which includes the PMI's Standards and Publishing departments and its Knowledge and Wisdom Center. He has sponsored and guided development and publishing of *A Guide to the Project Management Book of Knowledge* (PMBOK® Guide), 2000 edition; the *Practice Standard for Work Breakdown Structures*, the Government Extension to the PMBOK® Guide; the *Project Manager Competency Development Framework* Construction Extension to the PMBOK® Guide (Provisional Standard); and, most recently, the *Organizational Project Management Maturity Model* (OPM3). Prior to joining the PMI in July 2000 he had a 30-year career in the U.S. Navy, as a helicopter pilot and acquisition professional. While in the Navy, he was commanding officer of Helicopter Anti-Submarine Squadron Three and program manager of the U.S. Marine Corps Light/Attack Helicopter Program. Mr. Fahrenkrog has a B.A. in economics from Moorhead State College and M.S. degrees in aeronautical engineering and management from the Naval Postgraduate School.

David N. Ford is an assistant professor of civil engineering at Texas A&M University. His expertise is in evaluating project management with analytical methods and simulations. He researches the dynamics of project management and strategy of construction organizations and teaches project management and computer simulation courses. Current research projects include the investigation of causes of failures to implement fast-track processes and the value of contingent decisions in project strategies. Prior to his appointment at Texas A&M, Dr. Ford was an associate professor in the Department of Information Sciences at the University of Bergen in Norway. He was one of two professors to develop and lead the graduate program in the system dynamics methodology for 4 years. Dr. Ford's research during this time focused on the dynamics of product development processes and included work with Ericsson Microwave to improve its product development processes. Dr. Ford designed and managed the development and construction of facilities for over 14 years in professional practice for owners, design professionals, and builders. Projects varied in size and facility types, including commercial buildings, residential development, and industrial, commercial, and defense facilities. He serves as a reviewer for *Management Science*, the *Journal of Operational Research Society*, *Technology Studies*, and *System Dynamics Review*. Dr. Ford received his B.C.E. and M.E. from Tulane University and his Ph.D. from the Massachusetts Institute of Technology in dynamic engineering systems.

Theodore C. Kennedy (NAE) is chairman and co-founder of BE&K, a privately held international design-build firm that provides engineering, construction, and maintenance for process-oriented industries and commercial real estate projects. Mr. Kennedy's experience and expertise are in the design, construction, and cost estimation of complex construction and engineering projects. BE&K companies design and build for a variety of industries, including pulp and paper, chemical, oil and gas, steel, power, pharmaceutical, and food processing, among others. BE&K is consistently listed as one of *Fortune* magazine's Top 100 Companies to Work For, and BE&K and its subsidiaries have won numerous awards for excellence, innovation, and programs that support its workers and communities. Mr. Kennedy is the chairman of the national board of directors of INROADS, Inc., and is a member of numerous other boards, including the A+ Education Foundation and the Community Foundation of Greater Birmingham. Mr. Kennedy is also a member of the Duke University School of Engineering Dean's Council and former chair of the Board of Visitors for the Duke University School of Engineering. He is the former president of Associated Builders and Contractors and the former chairman of the Construction Industry Institute. He has received numerous awards, including the Distinguished Alumnus Award from Duke University, the Walter A. Nashert Constructor Award, the President's Award from the National Association of Women in Construction, and the Contractor of the Year award from the Associated Builders and Contractors. Mr. Kennedy has a B.S. in civil engineering from Duke University.

Stephen R. Thomas is an associate director of the Construction Industry Institute (CII) and director of the CII Benchmarking and Metrics Program. In this position he establishes and disseminates performance and practice use metrics for the construction industry and directs the development and operation of a research-based benchmarking system for capital projects improvement. Dr. Thomas is a member of the steering committee for the National Institute of Standards and Technology (NIST) project to develop a cost-effectiveness tool for evaluating the management of terrorist risks and completed several NIST-sponsored research projects from 1998 through 2004 as the principal investigator. He is also a senior lecturer in the University of Texas (UT) at Austin Department of Civil Engineering for both undergraduate and graduate courses in project management and analytical analysis. Prior to coming to UT he was an assistant professor in the United States Military Academy's Department of Systems Engineering and chief of engineering and construction management for a U.S. Army engineer brigade in Germany. He has coauthored numerous articles on construction and project management. He received a B.S. in engineering from the United States Military Academy and M.S. and Ph.D. degrees in engineering from the University of Texas at Austin.

D
Discussions with DOE Personnel

OCTOBER 27, 2004

Overview and Discussion of Current DOE Performance Measures and Benchmarking

- Current GPRA performance measures
- Internal benchmarking
- Project Assessment and Reporting System (PARS)
- Environmental Management Project Definition Rating Index
- Idaho National Engineering and Governmental Laboratory pilot evaluation
- DOE project management personnel data

Department of Energy Participants

Syed Bokhari, Office of Civilian Radioactive Waste
Saralyn Bunch, Office of Civilian Radioactive Waste
Kevin Clark, Office of the Assistant Secretary for Fossil Energy
Mike Donnelly, Office of Engineering and Construction Management
Marvin Gorelick, Office of the Assistant Secretary for Energy Efficiency and Renewable Energy
Lenny Mucciaro, Office of Environmental Management
Tom O'Connor, Office of Nuclear Energy Science and Technology
Dale Oliff, National Nuclear Security Administration
Dave Pepson, Office of Engineering and Construction Management

Other Participants

Amar Chakar, Civil Engineering Research Foundation
Marvin Oey, Civil Engineering Research Foundation

DECEMBER 15, 2004

Roundtable Discussion of Possible Project Management Performance Measures and Benchmarking Procedures

- DOE expectations and special issues
- Discussion of the relevance, ease of measurement, and usefulness of performance metrics
- Benchmarking procedures

Department of Energy Participants

Saralyn Bunch, Office of Civilian Radioactive Waste
Kevin Clark, Office of the Assistant Secretary for Fossil Energy
Mike Donnelly, Office of Engineering and Construction Management
Marvin Gorelick, Office of the Assistant Secretary for Energy Efficiency and Renewable Energy
Karen Guevara, Office of Environmental Management
Steve Meador, Office of Science
Richard Minning, Office of Program Management
Tom O'Connor, Office of Nuclear Energy Science and Technology
Dale Oliff, National Nuclear Security Administration
Dave Pepson, Office of Engineering and Construction Management
Jeffrey Short, Office of Policy and Site Transition

Other Participants

Amar Chakar, Civil Engineering Research Foundation